Treasures of the Earth

Rocks and Minerals Picture Book for Kids

David J Riffey

Sycamore

CONTENTS

WHERE DO ROCKS COME FROM?

The Earth is a big ball of rock! From the highest mountain to the center of the planet, everything is made of rocks. Even the plants, animals and all the people are made of stuff that came from rocks.

When you look at a picture of the Earth, most of it is water. And although we don't think of it because water is mostly liquid, water is a mineral. And minerals are a kind of rock.

Deep inside the Earth it is very hot! So hot that it is molten. The liquid rock is under tremendous pressure.

Gradually the inside of the Earth gets cooler and hardens in to rock.
This hardened rock is what makes the Earth's Crust. The crust is what we live on. All the mountains. All the oceans. All the land is on the top part of the Earth's crust.

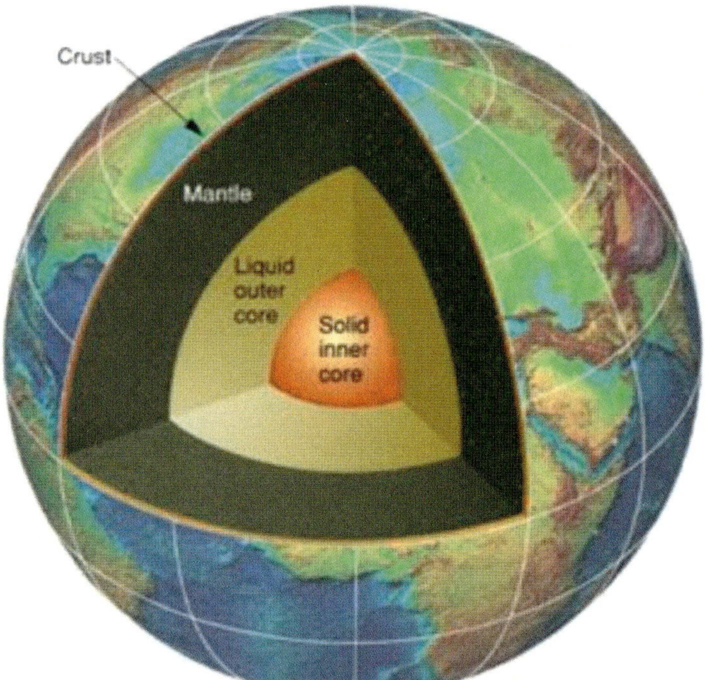

Kinds of Rock

There are basically three kinds of rock.

Igneous

Sedimentary

Metamorphic

How rock is formed.

Down beneath the Earth's crust is the next layer. The Mantle.
The Mantle is very hot. It contains liquid rock called Magma.

Magma gradually gets cooler as it reaches the surface of the crust. Eventually it hardens to become rock.

This is called Igneous rock.
Igneous means “of fire”.

Sometimes the liquid magma leaks out on to the surface of the Earth.
Then the magma is called lava.
When liquid lava comes out of the ground, it forms Volcanos.
The lava hardens into volcanic rock.

This is a lava flow in Hawaii. The liquid lava spews out and hardens into black rock. Eventually it can pile up and form a mountain. The islands of Hawaii are all formed by volcanic activity.

The lava is so hot that it can start fires. This picture shows a lava flow moving into a papaya tree orchard. If there were any houses in the way, they were destroyed too. But for all the danger, there is beauty too. The beautiful islands of Hawaii are formed from volcanos. The soil formed from the volcanos is great for growing plants. The Hawaiians have vast farms. So, despite the danger, people learn to live and thrive in volcanic regions.

Augustine Volcano viewed from the west.

This is a volcano in Alaska, USA.

Volcanic rock from Pico de Fogo (Cap-Vert)

**Pumice stones
from the beach of the island
of Stromboli, Sicily, Italy**

Sometimes the lava cools in small round balls and turns into volcanic glass. Volcanic glass is called obsidian.

Obsidian

The colors can range to deep black to brown. Because of the mineral content inside the glass, it is not clear like the glass we use in windows. The smaller black oval shaped ones in this picture are called “Apache's Tears”.

Sometimes a volcano will eject rock dust. Volcanic ash. This can range from a small dusting, to tons of ash piling up on the ground. Sometimes it can block out the sun. Sometime it can pile up and bury a village.

Volcanic Ash

Volcanic Ash Flow

Other kinds of igneous rocks are not dramatic as the kind made by volcanos. In some ways this other kind is more important. And I bet you have some around you now but didn't even know it.

This other kind of igneous rock is formed slowly deep within the crust of the earth. The hot liquid mantle seeps into the crust through cracks. Some time the liquid comes to the surface and forms a volcano. Sometimes the magma does not make it out, but just slowly cools. The rock gets hard taking many years to get to its final stage. Sometimes the ingredients in the rock separate from the rest and form into crystals. This kind of igneous rock is usually very hard and very heavy.

One of the most prominent igneous rocks is granite.

Granite

Granite is very hard and durable. It often used for building material. Flooring, countertops and even headstones in cemeteries are often made of granite. Granite can be cut, ground smooth, and polished to a beautiful shine.

Sedimentary Rock

As solid as rocks are, they are constantly being worn down by erosion. When rain, water, wind, ice, or sand move across rock they wear down to smaller pieces. Big mountains can break into small boulders. Boulders become rocks. Rocks become pebbles. Pebbles become sand.

The mountains are washed into rivers. Rivers wash the rocks into smooth rounded shapes. Eventually the rocks pebbles and sand settle down. They settle on the bottom of rivers, lakes and even oceans. After a long time, the sand, mud, and rocks and pebbles can harden into layers of rock.

Sedimentary rock that was formed under the deep water of an ocean or lake can be uplifted into a mountain. Then erosion wears away the top layers revealing the hard sedimentary rock beneath.

Limestone layers at the Grand Canyon
Grand Canyon National Park, Arizona. Looking toward Bralina Temple from top of Bright Angel trail.

The word Sedimentary come from the word sediment. Layers of sand and mud settles to the bottom to form strata.
You can see the flat layers or strata in the side of this mountain.

Grand Canyon Red Sandstone Strata

This picture is of another part of The Grand Canyon. Water has eroded the mountains creating a Canyon, through which a river flows. You can see the red layers of sandstone in these mountains.

When sedimentary rock forms in the ocean, it usually includes the shells of sea creatures that have died as the sediment was being deposited. The shells of mollusks like clams and snails are made of a mineral called calcium carbonate. Calcium carbonate is also known as lime. The sedimentary rock formed under the ocean is usually called limestone. It contains calcium carbonate left behind by sea creatures.

Limestone

Sometimes the layers of sediment are mostly sand. Like at the bottom of a large lake. The rock formed those cases is called sandstone. Sandstone can come in many colors but The Grand Canyon is famous for it's red sandstone.

Grey Sandstone

Carbonaceous shale

When the layers are made of mud, the resulting rock is called shale

Metamorphic Rock

The word metamorphic means change. Metamorphic rock is rock that has been changed from one kind or rock to another. Usually way down deep in the Earth's crust. The rock heats up but not quite to the point of melting. And the pressure of all the rock on top squeezes down. Then after many years the rock changes. Usually the rocks get harder, denser, and more finely grained.

You probably have metamorphic rocks around you and don't even know it.

Marble

One example of metamorphic rock is marble.

Marble was once limestone that was changed by heat and

pressure to a harder, finer rock.

Marble – pink

Mable comes in many colors. White, grey, pink. Usually it has black veins of minerals going through it. Highly prized for it's beauty. Marble is used in building material. Countertops, flooring, gravestones. carvings and statues.

Quartzite

If you take sandstone and give it enough time, heat and pressure it turns into quartzite. Unlike sandstone, which is easily scratched even by a fingernail, quartzite is hard and tough.

Kyanite Quartzite

Quartzite is usually grey or white in color. It has a glossy surface and is sleek. Wall covering, flooring, roofing tiles and even bricks can be made of quartzite. Crushed into stone, it is used in road construction.

Slate

So mud becomes shale.
If you give it the metamorphic process shale become slate.
Again, slate is denser, harder and more finely grained.

Slate

Slate tends to break easily into its layers. So it makes flat plates of smooth rock. So that is why slate is often used in chalkboards.
Slate is often a black color but can be gray or many other colors.
Because of it's hardness and durability, slate is used in floors, countertops, and even roof tiles.

ROCKS ARE USEFUL

Ok everything on the earth is basically made of rock. And rocks are everywhere. People need rocks. People need the stuff that rocks are made out of.

What are rocks used for? Everything!

We just saw how some kinds of rock can be cut and used for building material. Granite, marble, quartzite, for example.

Do your parents own a car? Do you have a bicycle. Do you take a bus to school? Those are made of metal. They may be made of steel that is made from iron, which is a metal that comes from rocks called iron ore. An ore is a rock made of a valuable mineral element such as iron and another chemical element such as sulfur. Miners dig them up in mines and quarries. Then the valuable part of the rock is extracted. Ores can contain different valuable elements. Gold ore contains gold. Copper ore contains copper. Silver ore contains silver.

The pumice that comes out of volcanos can be used as an abrasive. There is a popular hand soap that contains pumice. It is used to clean grease and grime off of hands.

Concrete Blocks

Concrete is used in building material. Concrete is a mixture of water, crushed rock, sand and cement that hardens into a man made stone.
Cement is made from limestone. Concrete can be poured out as a liquid into a flat slab. It can be poured into a mold to different shapes.
Concrete blocks are used to build houses. Concrete is used to make driveways and parking areas for cars.

Bricks in a pile

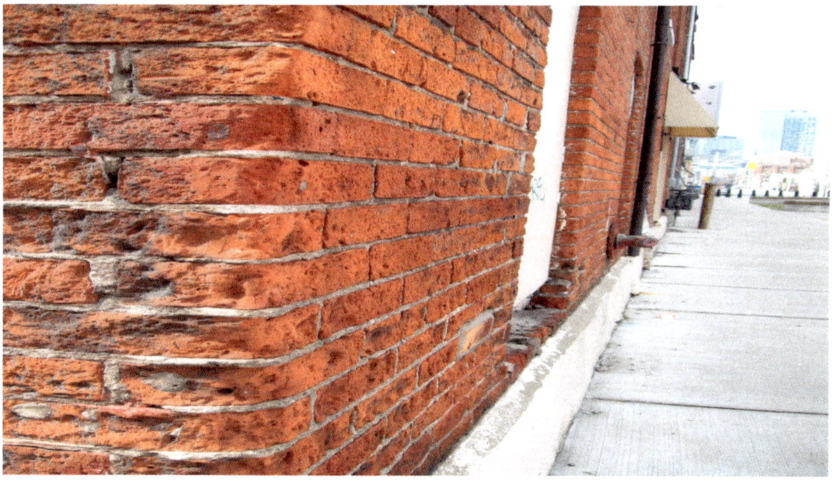

Brick wall

Bricks are a building material made from clay. Clay is mixed with water and other ingredients and molded into rectangles. They are heated until they become solid. Bricks are used in walls as well has walkways and roads.

US Navy personnel installing drywall in a home.

If the walls in your home are made of drywall, then you have rocks in your walls because drywall is a flat board made of a rock called gypsum. Drywall is also called sheetrock. It makes a nice smooth wall that is easy to install and paint. It is also fire resistant. That means it is safer than wood or paneling.

Gypsum in its raw form

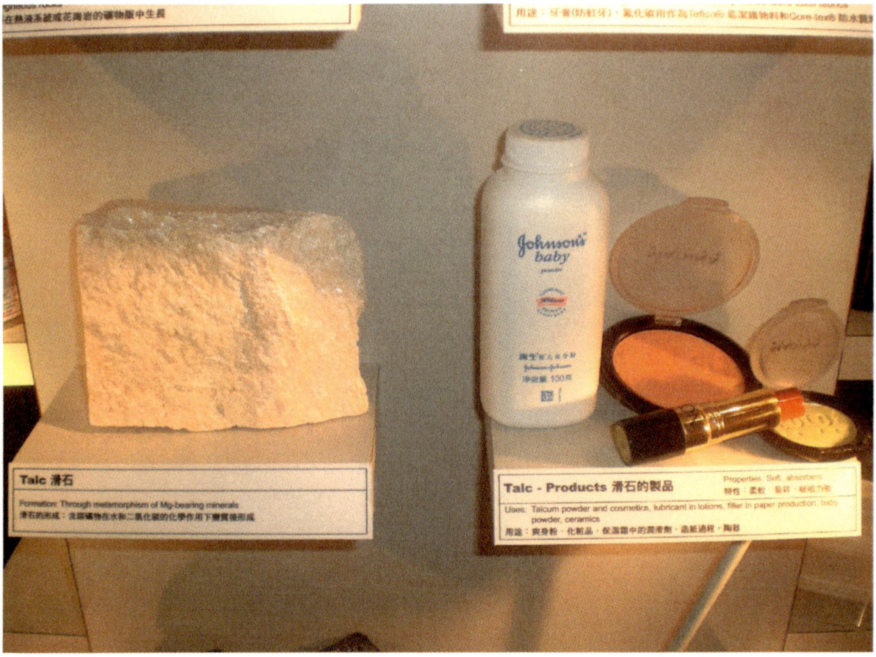

Talc and its uses.

One type of rock you may have around your house is talc.
Talc is a very soft mineral that is used in talcum powder. Talcum powder is sometime used in baby powder. Athletes sometime sprinkle it on their hands. Talc is used in cosmetics such as facial powders and lipsticks. Talcum powder makes skin feel smooth and dry.

Sea Salt

Most people never think about eating rocks. But we eat rocks every day. Salt is a mineral that is mined from deep underground. It can be extracted from sea water because salt is dissolved in it.

Salt in the rock form is called Halite. It is also known as table salt. Salt is used to season food to make it taste better. It was also used as a preservative before canning or refrigeration was invented.

Salt has been valued since the beginning of civilization. Salt was traded like money. The Roman Army was paid with salt. A good worker was said to be "Worth his salt". Salt is needed to keep people healthy, because it is an ingredient in our blood. So eating salt is good for you. But eating too much salt can be bad for you.

Coal and a US quarter used to show scale.

Coal is a rock that many people rely on but don't know it. Most people don't have coal in their homes anymore. Coal is a rock mostly made of carbon. It burns. It used to be a main source of heat in houses. Now, most people use oil, gas or electricity to provide warmth in the house. Coal is often used in electrical generation plants. The heat is used to produce steam that turns the generator and produces electricity. So even if you don't have coal in you own house, chances are you get your electricity from it.

Iron Ore with pen for scale.

Iron is metallic element that is used to make steel. Iron is very strong. Before steel was developed, iron was used to make tools and weapons. Also some buildings and bridges were made of iron. Iron is strong but it is also fragile. That means that it can crack. That is why steel is used instead. Steel is stronger and more flexible. Girders in bridges and buildings are made of steel.

Steel girders create sturdy framing for buildings

Copper

Copper is a metallic element. It usually has a shiny reddish appearance. It is easily made into things because it is soft. It can be pounded into shape or melted and poured into a mold. It is used in water pipes as well as air conditioning tubes. And it can be stamped into coins.

Copper coins

Carbon

Carbon is a element that is very useful. In it's natural form, it is black in color. It may be shiny or dull in appearance. It is used to make pencil leads. The lead in pencils is not the metal lead, it is graphite. Pure carbon. Graphite is soft and slippery when ground into a power. Graphite is used to lubricate locks so they operate better

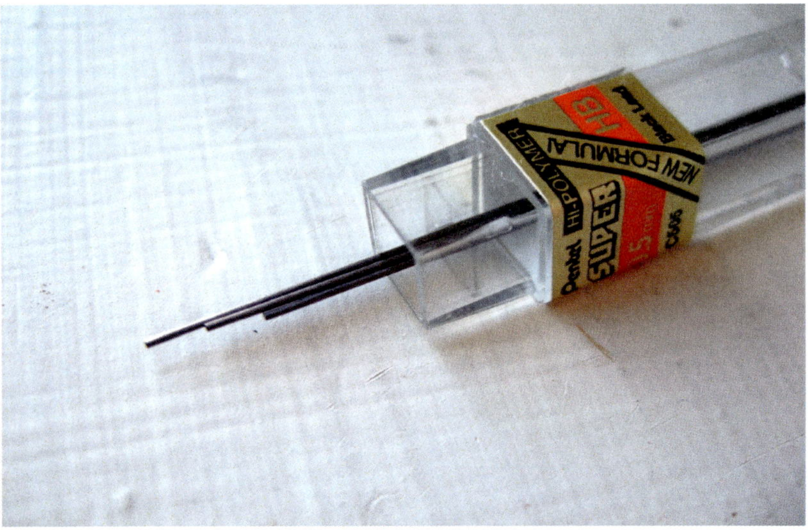

Graphite – pencil leads

Uncut Diamond

Diamonds Brilliant Cut

The amazing thing about carbon is that depending on the conditions it is formed in, carbon can make graphite or diamonds. Graphite is so soft that it can be scratched by a fingernail. If you rub graphite on a piece of paper, it will leave a dark streak. That is why it is used in pencils. Diamond is the crystallized form of carbon. Diamonds are so hard that they can only be scratched by other diamonds. If they are clear without any internal flaws, diamonds can be cut and polished to make beautiful jewelry.

Silver

Silver is a white shiny metal. It is used to make jewelry and knives forks and spoons for tableware. It is used in chemical and industrial processes. It is used to make coins. Because of the value of silver, it is purchased by investors.

Gold Nugget

Gold is a yellow shiny metal. Prized for it's beauty and rarity, it is made into jewelry. It is soft and is easily formed into shapes by beating with a hammer. Gold can be melted and poured into molds. Gold is also made into gold coins. Pirates were known to steal doubloons and "pieces of eight". When gold is made into coins it is easier to use for money than a big chunk of metal like this picture.

ROCKS ARE COOL

Many years ago, my Uncle gave me a box of rocks. It was a old wooden silverware box with some things he had collected over the years. Inside the box were some quartz crystals. A couple pieces of "fools gold". A couple arrowheads. Even a larger spearhead. From those items, I was inspired to learn more about them. I got a couple books on rock collecting. I became a "rockhound". A rockhound is someone who collects rocks and minerals as a hobby. My parents took me to local rock shops where I added to my collection as I could afford to. We even took a few classes in lapidary, which is polishing rocks into gemstones for jewelry.

I really like things that are beautiful or unusual.

Crystals are my favorite.

When minerals form under the right conditions, the molecules organize themselves in a pattern called a crystal. The same kind of mineral tends to make the same kind of crystal. The crystals can vary in color. Some crystals can be clearer than others. Some can include imperfections called inclusions. Sometimes crystals can grow to be very big. There are collectors all around the world.

Quartz Crystals

Quartz is a mineral made of the elements silicon and oxygen. Silicon Dioxide. One atom of silicon and two atoms of oxygen combine to make this chemical compound. Quartz is one of the most abundance minerals in the world. Most of the sand in the world is made up of quartz. The rocks marble and quartzite is almost completely made of quartz. And quartz is a major ingredient in granite.

Under the right conditions, quartz will grow into six sided or hexagonal shaped crystals.

Quartz can come in many colors.

When it is clear it is called a rock crystal.

Smokey Quartz

Sometimes another mineral gets mixed in when the crystal forms. That gives the crystal a color. Smokey quartz is clear with a brown to grey color.

Rutilated Quartz

Sometimes another mineral forms crystals inside the quartz crystal. Rutilated quartz has thin crystals of rutile inside the quartz crystal like tiny hairs. Commonly, rutile comes in reddish brown, but also in yellow, blue and violet. Rarely, it can be found in green or black.

Amethyst

When quartz crystals are purple they are called amethyst. The amethyst purple color comes from iron inside the crystal. It can also be caused by exposure to radiation when deep inside the earth. There may have been a radioactive rock like uranium ore near by, which affected the color of the crystals.

Amethyst is prized for beauty and hardness.

Clear flawless amethyst crystals can be made into gemstones and set into jewelry like rings or pendants.

Pyrite

Pyrite is a shiny yellow colored metal. Gold prospector would sometimes mistake it for gold. They called it “fools gold”.

Pyrite is a compound of iron and sulfur.
It is used in electronics and batteries.
When struck against steel, it produces sparks. Pyrite was used in early firearms that used a “wheelhouse” to ignite the gunpowder.

Pyrite is too brittle to be made into jewelry. If you hit with a hammer or drop it on the floor it will break into pieces.

Fluorite

Fluorite is a mineral made of calcium and fluorine. CaF_2. It comes in many different colors. It is softer than glass. Fluorite crystals can be cubic or double pyramid shapes.

Fluorite is used in industrial purposes like steel making. It is used to make high quality lenses in telescopes and microscopes.

Because it easy to scratch, it is seldom used in jewelry.

Obsidian with Red Streaks
Volcanic glass

When lava comes out a volcano, if it cools very fast it doesn't have time to form crystals. It turns to glass. The volcanic glass is usually a dark color, brown or black. Because of the minerals inside, it is not clear like window glass. This one happens to be shiny and has reddish streaks.

Obsidian can be cut and polished into gemstones, but is very soft and can break like window glass.

It does not hold up like harder gemstones.

Calcite Crystals

Calcite is a mineral made of calcium, carbon and oxygen. CaCO2. It comes in many colors. White or colorless, yellow, orange, even green. The colors come from impurities mixed in when the crystals formed.

Calcite is soft. Easily scratched. Even though it is pretty and shiny, it can not be used for gemstones. It can be dissolved by acids. Even a weak acid like vinegar can eat and damage the crystals.

Calcite has unique optical properties. A very clear crystal, like the one in the picture has double refraction. When something is viewed through it, the picture is doubled. Calcite was used in range finders. A rangefinder tells the viewer how far away something is, with out having to measure it out. Very handy to know if you are shooting at something. Calcite was used in gun-sites during World War Two.

Aquamarine

Aquamarine is a blue green gemstone. It is a form of Beryl.
Beryl is made of beryllium, aluminum, silicon and oxygen.
It can be colorless, green, blue, yellow, red, or white.

Gemstones are given different names when they are different colors. If beryl is green it is called emerald.

Aquamarine, emeralds and beryl are very hard and shiny.

They make great gemstones.

Very clear, large ones are very valuable.

Ruby

Ruby is a gemstone of the mineral corundum. Red in color and very hard. Clear rubies can be made into beautiful polished gems.

If it is another color than red, it is called a sapphire.

Rubies, if clear, are usually cut into faceted stones like a diamond. Sapphires are cut into dome shaped stones called cabochons. When cut and polished a sapphire may show a star shaped light reflection on the top center of the stone. That is a star sapphire.

If the crystals are very small and poor quality, the can be used to sand and polish other stones. They are made into sandpaper. The corundum grit in sand paper is sometimes call emery. The color of the grit may be reddish brown or black.

Ruby
Facet cut and polished

Star Sapphire
Cabochon with six point star

Sycamore

ISBN-13: 978-0615847078 (Sycamore Publishing)

ISBN-10: 0615847072

PHOTO CREDITS

Amythyst

Amethyst Améthyste2_(Brésil)

http://commons.wikimedia.org/wiki/File:Améthyste2_(Brésil).jpg

Amethystkristalle 08/12/08 Parent Gary

Earth blue marble

Image Credit: NASA/NOAA/GSFC/Suomi NPP/VIIRS/Norman Kuring

Earthcrosssection

http://www.nersc.gov/news-publications/news/science-news/2010/simulations-reveal-that-earth-ssilica-is-predominantly-superficial

Pahoeoe Fountain

http://upload.wikimedia.org/wikipedia/commons/9/9f/Pahoeoe_fountain_edit2.jpg

This image is in the public domain because it contains materials that originally came from the United States Geological Survey, an agency of the United States Department of the Interior. For more information, see the official USGS copyright policy.

volcanic eruption hvo00032

Hawaii Volcanoes National Park. 1960 eruption of Kilauea Volcano. Papaya and fountain at Kapoho. January 1960.

Image file: /htmlorg/lpb037/land/hvo00032.jpg

http://libraryphoto.cr.usgs.gov/cgi-bin/show_picture.cgi?ID=ID.%20HVO%20%2032ct&SIZE=large

Augustine Volcano viewed from the west.

Alaska Volcano Observatory (http://www.avo.alaska.edu). Image courtesy of U.S. Geological Survey. Image Creator: McGimsey, Game.

URL: http://www.avo.alaska.edu/image_full.php?id=5927

Roche volcanique au Pico de Fogo (Cap-Vert)

23 February 2012 Author Ji-Elle

Pumice

Pumice stones from the beach of the island of Stromboli, Sicily, Italy 18 April 2011 Norbert Nagel, Mörfelden-Walldorf, Germany

Object location 38° 48. 26.46. N, 15° 14. 6.25. E

Obsidian group

Author's personal collection

Volcanic ash

VolcanicAshUSGOV.jpg (360 . 246 pixels, file size: 36 KB, MIME type: image/jpeg)

Volcanic ash

Source:http://resourcescommittee.house.gov/subcommittees/emr/usgsweb/photogallery/ ; English Wikipedia, original upload 3 August 2004 by Chris 73

This image is a work of the "Minerals in Your World" project, a cooperative effort between the United States Geological Survey and the Mineral Information Institute. The images were featured in the "Minerals and Materials Photo Gallery" on the website of the U.S House Subcommittee on Energy and Natural Resources. As a work of the U.S. federal government, the image is in the public domain.

Pyroclastic flows at Mayon Volcano, Philippines, 1984.

Pyroclastic flows descend the south-eastern flank of Mayon Volcano, Philippines. Maximum height of the eruption column was 15 km above sea level, and volcanic ash fell within about 50 km toward the west. There were no casualties from the 1984 eruption because more than 73,000 people evacuated the danger zones as recommended by scientists of the Philippine Institute of Volcanology and Seismology.

http://volcanoes.usgs.gov/Imgs/Jpg/Mayon/32923351-020_caption.html

C.G. Newhall Derivative works of this file: Pyroclastic flows at Mayon Volcano-2010-20-08.jpg

Granite

http://marsrovers.nasa.gov/classroom/schoolhouse/rocklibrary/source/alkalic_granite.html

Grand Canyon crb00660

Grand Canyon National Park, Arizona. Looking toward Bralina Temple from top of Bright Angel trail. May 16, 1953

Image file: /htmlorg/lpb338/land/crb00660.jpg

http://libraryphoto.cr.usgs.gov/cgi-in/show_picture.cgi?ID=ID.%20Colton,%20R.B.
%20660ct&SIZE=large

Grand Canyon National Park, Arizona. Across the Grand Canyon from Hermits Rest. September 4, 1983.

Image file: /htmllib/batch92/batch92j/batch92z/batch92/eca00014.jpg

http://libraryphoto.cr.usgs.gov/cgi-bin/show_picture.cgi?ID=ID.%20Edwards,%20C.A.

%2014ct&SIZE=medium

limestone

http://marsrovers.nasa.gov/classroom/schoolhouse/rocklibrary/source/dolomitic_limestone.html

sandstone

http://marsrovers.nasa.gov/classroom/schoolhouse/rocklibrary/source/gray_sandstone.html

shale

http://marsrovers.nasa.gov/classroom/schoolhouse/rocklibrary/source/carbonaceous_shale.html

marble

http://marsrovers.nasa.gov/classroom/schoolhouse/rocklibrary/source/marble-pink.html

quartzite

http://marsrovers.nasa.gov/classroom/schoolhouse/rocklibrary/source/kyanite_quartzite.html

Slate

http://marsrovers.nasa.gov/classroom/schoolhouse/rocklibrary/source/slate-gray.html

Talc

http://commons.wikimedia.org/wiki/File:HKU_Stephen_Hui_Museum_Rock_Talc_..
_Johnson%27s_Baby_Powder.JPG

zh:............ Category:Stephen Hui Geological Museum Category:Stone samples Category:Exhibits in Stephen HuiGeological Museum zh:.. Category:Exhibits in Hong Kong [[Cate

Salt

Fleur de Sel sea salt from Guérande, France 18 April 2007

Christian Mertes (Muddl 12:26, 18 April 2007 (UTC))

Concrete block

http://www.dnr.state.oh.us/wildlife/Home/fishing/pond/fishproduction/tabid/6223/Default.aspx

Bricks

http://upload.wikimedia.org/wikipedia/commons/9/92/Weathered_bricks%2C_distillery_district%2C
_Toronto.jpg

Weathered bricks, distillery district, Toronto.

http://commons.wikimedia.org/wiki/File:Weathered_bricks,_distillery_district,_Toronto.jpg

Geo Swan9 May 2009 (talk)

Drywall

http://commons.wikimedia.org/wiki/File:US_Navy_031013-N-8208G002_Sailors_from_the_guided_missile_frigate_USS_Thach_(FFG_43)_move_drywall_for_Habitat_For_Humanity_International_during_San_Francisco_Fleet_Week.jpg

English: San Francisco Bay (Oct. 13, 2003) -- Sailors from the guided missile frigate USS Thach (FFG 43) move drywall for Habitat For Humanity International during San Francisco Fleet Week. Habitat for Humanity International is a nonprofit ecumenical Christian housing organization that helps build simple, decent, affordable houses in partnership with those in need of adequate shelter. The celebration is held at the Fisherman's Wharf featuring 2,500 Sailors, Navy Blue Angels and U.S. Navy ships; USS Cleveland (LPD 7),

USS Hopper (DDG 70), USS Thach (FFG 43), USS Shiloh (CG 67), USS Mobile Bay (CG 53). U.S. Navy photo by Aviation Electronics Technician Airman Sandra Gomes. (RELEASED) 13 October 2003 This Image was released by the United States Navy with the ID 031013-N-8208G-002 (next). This tag does not indicate the copyright status of the attached work. A normal copyright tag is still required. See Commons:Licensing for more information.

U.S. Navy photo by Aviation Electronics Technician Airman Sandra Gomes.

gypsum

http://commons.wikimedia.org/wiki/File:Gypsum_cones.jpg

http://upload.wikimedia.org/wikipedia/commons/9/9b/Gypsum_cones.jpg

Gypsum in it's raw form

Gypsum of the Yesares member of the Sorbas basin, formed in association with the Messinian salinity crisis. The gypsum can be clearly seen to have formed at the sediment surface, forming cones separated by laminated sediments.

Verisimilus

coal byu00329

Coal (Cannal). Quarter for scale. Mineral collection of Brigham Young University Department of Geology, Provo, Utah. Photograph by Andrew Silver. BYU index 1-1013b.

Image file: /htmlorg/lpb553/land/byu00329.jpg

http://libraryphoto.cr.usgs.gov/cgi-bin/show_picture.cgi?ID=ID.%20BYU%20Mineral

%20Specimens%20329&SIZE=large

iron ore byu00672

Iron (terrestrial). Pen for scale. Mineral collection of Brigham Young University Department of

Geology, Provo, Utah. Photograph by Andrew Silver. BYU index 1-4079a.

Image file: /htmllib/btch567/btch567j/btch567z/byu00672.jpg

http://libraryphoto.cr.usgs.gov/cgi-bin/show_picture.cgi?ID=ID.%20BYU%20Mineral
%20Specimens%206

steel support beam

http://www.photos-public-domain.com/wp-content/uploads/2010/12/steel_support_beam_girder.jpg

Bricks pile

Attribution © <A HREF="http://taluda.openphoto.net">Adrian van Leen</A> for <A

HREF="http://21233.openphoto.net">openphoto.net</A>

bricks

http://upload.wikimedia.org/wikipedia/commons/9/92/Weathered_bricks
%2C_distillery_district %2C_Toronto.jpg

COPPER

http://commons.wikimedia.org/wiki/File:Copper.jpg

copper coins

http://www.pdphoto.org/PictureDetail.php?mat=&pg=5426

graphite

http://commons.wikimedia.org/wiki/File:Min_graphite.jpg

pencil lead

Mechanical pencil leads spilling out of their plastic case 5/19/07 Mrs Scarborough

silver bom00011

Native silver. Konigsberg, Norway. Bureau of Mines, Mineral Specimens C\01640.

Image file: /htmlorg/lpb555/port/bom00011.jpg

http://libraryphoto.cr.usgs.gov/cgi-bin/show_picture.cgi?ID=ID.%20BOM%20Mineral

%20Specimens%20011&SIZE=large

Uncut Diamond

Uncut industrial-grade diamond. 31 August 2008 Shieldforyoureyes Dave Fischer

Diamonds Brilliant Cut de:Diamanten en:Diamond

4 February 2004 Mario Sarto Derivative works of this file: Cardinal gems.png

gold bom00031

Gold. Francomb Hill, Summit County, Colorado. Bureau of Mines, Mineral Specimens C\01683.

Image file: /htmllib/btch555/btch555j/btch555z/btch555/bom00031.jpg

http://libraryphoto.cr.usgs.gov/cgi-bin/show_picture.cgi?ID=ID.%20BOM%20Mineral
%20Specimens%20031&SIZE=medium

Quartz Crystal Group

Smokey Quartz

Rutilated Quartz

Amethyst Group

Pyrite Group

Fluorite Group

Obsidian with Red Streaks Volcanic Glass

Calcite pair

Aquamarine closeup

Ruby closeup

Author's personal collection David J Riffey 2013

Ruby facet cut and polished

ruby; red corundum 27 March 2009 Decym92

http://commons.wikimedia.org/wiki/File:Rubin013.JPG

star sapphire

http://commons.wikimedia.org/wiki/File:Star-Saphire.jpg

ABOUT THE AUTHOR

David J Riffey became a rockhound when he was a young boy.
His uncle gave him a old wooden silverware box filled with treasure.
Quartz crystals, fools gold, arrowheads and more.
That was the start of a lifelong fascination with rocks and geology.

He is a freelance writer living in St. Petersburg, Florida.
His parents moved down to Florida from the bitter cold of the north when he was a small child. Growing up in St. Petersburg he witnessed many changes, as the sleepy southern town became a thriving metro area.

In 1976 he graduated Sigma Cum Laude with a Bachelor's of Religious Education from Florida Beacon College in Largo, Florida

A varied career path has provided him with unique insights about business, investing, entrepreneurship, sales and marketing.

He has written and published fiction and non-fiction books on Amazon CreateSpace and Amazon Kindle.

OTHER SUGGESTED BOOKS

Here are a couple books on Amazon that have information about rocks and minerals. Please ask your parents before going to Amazon and buying them. You can look them up by the title on Amazon. I have also provided a short link to the page, if you want to use it.

http://goo.gl/ccGzVl

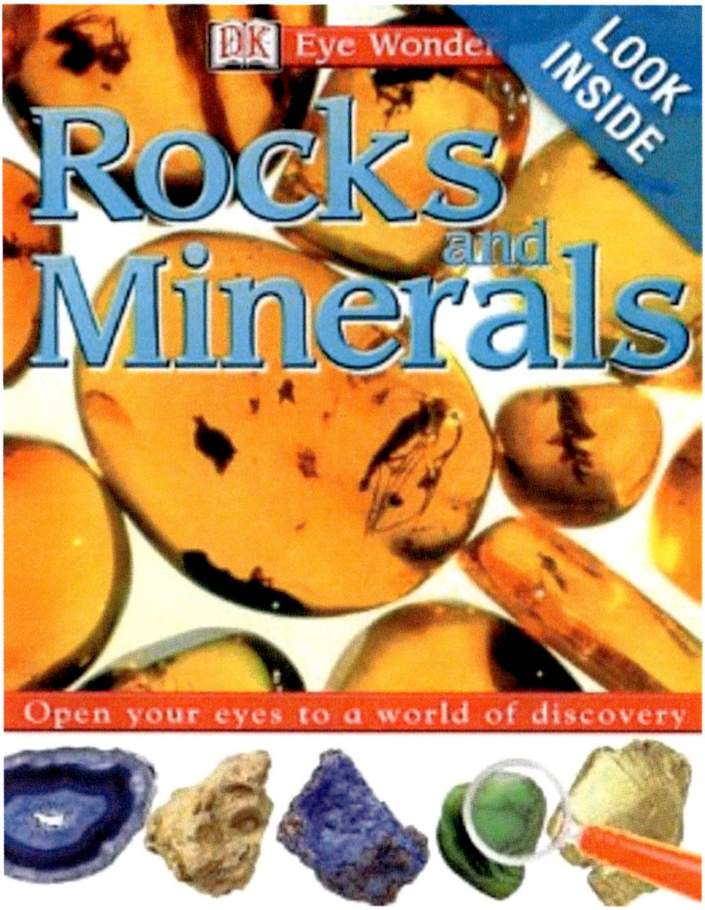

http://goo.gl/q5SyKR

American Educational Classroom Collection of Rocks and Minerals

http://goo.gl/F6HLPZ

Made in United States
Orlando, FL
26 October 2025

71372640R00045